DUST IN THE LONDON UNDERGROUND

A report by London Transport

C

Fo

Introduction *2*

Sources of dust *2*

Distribution of dust *3*

Composition of dust *5*

Asbestos *6*

Quartz *7*

Research methods *12*

Further action *13*

Collaboration and communication *15*

Summary *17*

Supporting information *18*

London: Her Majesty's Stationery Office

© *Crown copyright 1982*
First published January 1982

Enquiries regarding this publication should be addressed to Area Offices of the Health and Safety Executive, or the Public Enquiry Point, Baynards House, 1 Chepstow Place, London W2 4TF, tel 01-229 3456

ISBN 0 11 883622 6

Foreword

In 1967 it came to the attention of the London Transport Executive by complaints from members of the staff, the general public and press reports that the quantity of dust in the London Underground was creating discomfort and possibly a hazard to the health of those who travelled or worked in the Underground system.

As a result an extensive investigation was carried out by London Transport and its scientific and medical staff, with the assistance of those bodies listed on Page 15 of this report. The investigation included dust counting, dust analysis, research into the composition of brake blocks, research into operation of the trains, and analysis of the sickness records and X-rays of staff. A detailed technical report was written by the scientists involved and made available by London Transport to the Health and Safety Commission's Railways Industry Advisory Committee, of which I am Chairman.

In view of the possible wider interest in the subject, the Committee asked London Transport to prepare this report, summarising the findings of their research, in a form suitable for publication.

The concern of the report is with the possible effects on unprotected persons such as train crews, station staff and the travelling public; it does not cover the risks which may be encountered by contractors who have to remove and replace the sound-proofing insulation on tunnel walls nor staff of London Transport such as 'fluffers', who have to remove accumulated dust from the track bed and rails of the system, who are provided with appropriate respiratory protection.

The report has been considered by the Railways Industry Advisory Committee, which has noted the improvements effected by London Transport and the additional action they are proposing to take and has now recommended that the report be published for the information of staff and the travelling public, to reassure them that there is now no significant hazard from the levels of dust in the tunnels and stations of the London Underground.

Lt Col I K A McNAUGHTON
Chief Inspecting Officer of Railways

December 1980

Introduction

Dust has always been regarded as a problem on the London Underground, partly because concentrations of dust can cause unpleasant working conditions for staff and unpleasant travelling conditions for the public, and partly because it has been feared that some kinds of dust might cause actual damage to the health of staff and passengers.

So far as the effects of dust on the general working and travelling environment are concerned, the main factor is the total concentration of dust of all kinds in the air. From the amenity standpoint special attention has therefore been given to obvious dust 'black spots' such as Highgate, where there were clearly visible clouds of dust caused by the severe braking of trains on the down gradient into the station.

As regards the possibility of actual damage to health from Underground dust, the main factors are:

(a) the concentration of 'respirable' dust in the air, in other words, the proportion of the total dust which can actually be breathed into the lungs and retained there;

(b) the proportions of certain particular substances in the respirable dust which might cause health hazards if they exceeded accepted levels. These substance are asbestos and silica (of the type known as quartz);

(c) the duration of exposure to the dust.

Research and investigation into all the various aspects of the dust problem have been carried out by London Transport over many years, in collaboration with independent experts and professional bodies. This report sets out the results of this research and the conclusions reached, gives details of action already taken, and explains the further investigations and developments which are in hand or proposed.

Sources of dust

There are several sources of dust in the Underground, including:

(a) the wear on brake blocks and wheels when trains are being slowed down or stopped;

(b) the wear on the rims and flanges of wheels and on the heads of the rails, caused by the contact between the moving wheels and the track;

(c) the wear on other parts of the rolling stock, including the upholstery;

(d) the gradual wearing down of ballast and other civil engineering materials;
(e) oily matter from lubricating systems;
(f) wear on clothing;
(g) particles brought in from outside the Underground, either on people's clothes and footwear or through the ventilation system.

Of these different sources the first two, brake block and wheel/rail wear, are by far the most important.

Distribution of dust

There are strong air currents in the Underground and these largely dictate where the dust will collect and settle. Sometimes the effects are unexpected; for example, one part of a tunnel may be quite clean while an adjoining part is unduly dirty, owing to some quirk of the prevailing air currents in the area.

Without being too precise, it can be said generally that higher amounts of dust are found where trains are regularly subjected to heavy braking, as on downhill approaches to stations, and where trains have to negotiate sharp curves (thereby creating greater friction between wheel and rail). Frequently conditions in these areas where high concentrations of dust occur have been the subject of complaints by the staff and the public, but they can be greatly improved when remedial action is taken. The clearest case is that of Highgate, where as already mentioned, trains were being severely braked at the bottom of the long down gradient into the station. In 1967 the amounts of dust being generated were regarded by London Transport as unacceptable from the point of view of the well-being of staff and passengers, and in order to reduce them steps were taken to enforce speed restrictions and so cut down to a minimum the degree of braking required. The result was a dramatic improvement, and further benefits were found later following improvements in the manufacture of brake blocks to ensure more regular wear and the introduction of a proportion of new rolling stock with partial 'rheostatic' (i.e. non-friction) braking. The 'before and after' picture at Highgate is best shown in Fig. 1.

By comparison, samples taken at 31 other sites at 20 Underground stations in the summer and autumn of 1975 showed total dust levels varying from only 0.2 mg/m^3 at the street entrance of Holland Park Station to a little over 3 mg/m^3 on the southbound platform at Trafalgar Square Station. In no case were there any results remotely approaching the original severe dust levels at Highgate.

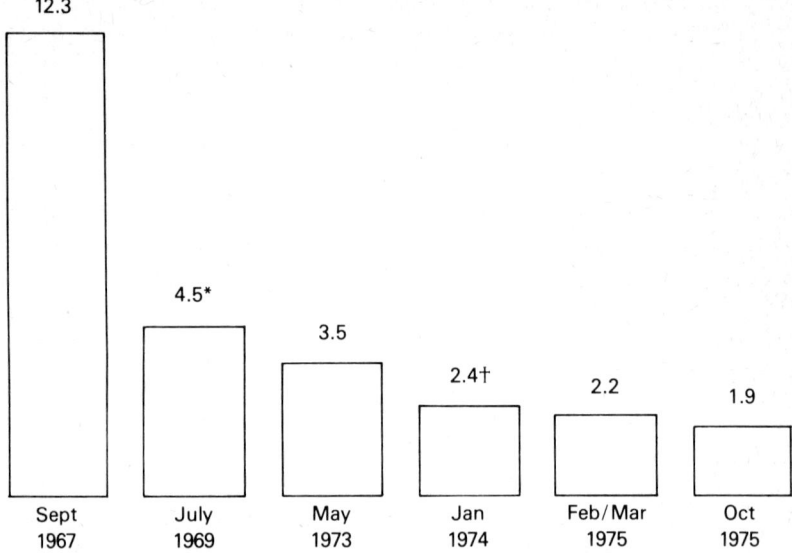

Highgate Station (Northern Line) — Southbound platform
Total dust in the air — average concentration
Figures show milligrammes per cubic metre (mg/m^3)

* Following enforcement of speed restrictions
†Following improved manufacture of brake blocks and introduction of some trains with partial rheostatic braking

Note September 1967 figures were above the threshold limit value (TLV) for nuisance dust (10 mg/m^3, total dust)

Fig 1

Another 'black spot' for excessive airborne dust was Baker Street (Bakerloo Line); conditions were somewhat abnormal because of construction work associated with the new Jubilee Line, but tests carried out before and after the introduction of special 'speed control' signalling showed that between May 1976 and August 1977 the level of airborne dust had been cut from 6.4 mg/m^3 to 2.2 mg/m^3.

So far as differences in dust levels between one Underground line and another are concerned, tests carried out at crew positions on trains running between November 1976 and August 1977 showed levels of respirable dust in the air varying between 1.0 mg/m^3 in the Victoria Line operator's cab and 2.4 mg/m^3 in the Bakerloo Line guard's position, the low figure for the Victoria Line being evidently mainly due to the rheostatic braking on all trains on that line. Other information obtained from these particular 'on-train' tests is considered later in this report.

It will be noted that all the comparisons made are on the basis of airborne dust. This is because dust which has already settled is not inhaled and presents no particular health hazards unless it again becomes airborne. Settled dust is not free of *all* risks (for example, smouldering) and it can be a nuisance to staff who, for one reason or another, may have to disturb or remove it.

Composition of dust

Over the past thirty years or more many chemical analyses have been made of Underground dust, either of actual airborne dust or of dust taken from cables and ledges where it had been deposited after being airborne. The analyses were made in connection with a variety of problems, including those of dirt on trains, risks of smouldering, and the design of a tunnel cleaning train which London Transport has been developing in recent years. The general make up of the dust is therefore well understood, though the proportions of some of its constituents may vary quite considerably.

The main feature of this Underground dust which makes it different from other kinds of dust normally encountered, is its high iron content (either as metal particles or as oxides), often amounting to more that 50% of the total. This is not unexpected in view of the main sources of Underground dust (wheels, rails and brake blocks), as described earlier. This iron dust is dirty, but it is generally accepted that there are no significant hazards in inhaling particles of iron or its oxides.

Another constituent of tunnel dust is silica, which is found both in the form of quartz and in combination with other elements. This is also to be expected, since brake blocks contain a proportion of quartz. Because it is recognised as a health hazard to inhale large amounts of quartz dust over a prolonged period, London Transport decided, wholly on its own initiative, to put a considerable amount of research effort into studying the quartz in the tunnel dust and its possible medical effects.

With the exception of asbestos, none of the remaining constituents of the Underground dust, i.e. moisture, clothing fibres and so on, is recognised as giving rise to any health hazard.

With regard to asbestos, the proportions present in the tunnel dust are so very small that it has been difficult even to measure them. Nevertheless, because of the known health risks involved in prolonged exposure to asbestos dust, this too has been the subject of careful investigation.

The following sections of the report deal respectively with the asbestos and quartz studies and their results.

Asbestos

There are three main types of asbestos, known generally as 'blue', 'white' and 'brown' asbestos. The dust from blue asbestos is considerably more dangerous than that from the white variety, but 90% of the asbestos produced throughout the world is of the white type.

The only known site in the Underground tunnels where blue asbestos was installed in substantial quantites (in the form of noise reducing coating) was the so-called 'Bull and Bush' site in the southbound tunnel of the Northern Line between Golders Green and Hampstead, and although London Transport was advised that this installation involved a very low order of risk as it stood, it was nevertheless decided to remove it. This was done by contractors specialising in such work, under the strictest safety conditions (involving temporary closure and sealing of the tunnel), in July 1978. It was also found that six tube cars had had blue asbestos noise-insulating material installed in them experimentally, and this material also has since been removed, with all the required safety precautions. Additionally some ventilation shafts were found to contain blue asbestos, which has now also been removed. So far as white asbestos is concerned, a survey showed a number of places where it is embodied in noise-reducing panels, waterproof tunnel linings, plasterboard facings, fire-resistant panels and cable coverings; but this asbestos is usually bonded or sealed, or covered by tunnel dust, and hazards could arise only on its installation or removal, or in the event of its being damaged.

The only other known source of asbestos in the Underground dust is the 5% of white asbestos in the brake blocks, and it would appear that such airborne asbestos as has been found comes mainly from there. Even then, however, it is an established fact that, during the braking operation, almost all the asbestos affected is converted by the heat generated into other substances which present no hazard, and the proportion of asbestos which survives to be released in the dust from the brake block wear is extremely small.

It should incidentally be noted that the 5% of asbestos in Underground brake blocks is very much less than the proportion found in brake blocks used by many other railway undertakings and also in brake linings used on road vehicles, which frequently contain more than 50% of asbestos.

Despite the generally reassuring picture, it was considered that there must be full investigation into the level of asbestos in Underground dust, to satisfy all concerned that it was of a negligible order. Tests by the London School of Hygiene in 1953, and again in 1967, using the best techniques available at the time, had shown that only a small fraction of 1% of the dust could be identified as asbestos.

Under statutory requirements occupational exposure to any form of asbestos dust should be reduced to the minimum that is reasonably practicable. In any case, exposure to blue asbestos should not exceed 0.2 fibres/ml for any 10-minute period and for other types of asbestos should not exceed 2.0 fibres/ml over a 4-hour period and/or 12 fibres/ml over any 10-minute period.

By the early 1970s, London Transport's own scientific tests had shown that the levels of asbestos in the air were very much lower than those limits, and in a joint report of 1974, based on tests using very advanced methods, (see page 12, last paragraph) the Asbestosis Research Council and the TUC Centenary Institute of Occupational Health were able to show that the asbestos concentrations in Underground tunnel air were in fact very much lower still, being between one-thousandth and one ten-thousandth of the 2 fibre/ml level for non-blue asbestos.

Later investigations by a total of five independent bodies have again shown extremely low levels. Tests made at sites where work on installations containing asbestos has been going on (e.g. where cables were being installed at Baker Street and where asbestos had been removed at the 'Bull and Bush' site) have shown that outside the actual working areas levels of asbestos in the air were far below the limit set in the hygiene standard.

Altogether there have been 24 series of airborne asbestos measurements at 12 London Transport stations over the years, and all the results confirm that there is no significant risk to the health of staff or passengers from asbestos in dust in the Underground.

Quartz

Reference was made earlier in this report to 'respirable' dust, and this is especially relevant to the question of the quartz particles found in the tunnel air. Whereas a proportion of the dust particles may be trapped in the nose and throat, those inhaled particles of a size capable of being retained in the terminal airspaces of the lungs are termed 'respirable' particles. The dust samples taken at a variety of Underground stations in 1975 indicated that between one-half and two-thirds of the dust was in the respirable range.

Because quartz forms a sizeable proportion of the total composition of standard Underground brake blocks, quartz is an important component of tunnel dust, and the tests at stations showed that at a few of the sites (and particularly at Highgate) the concentration of respirable quartz in the dust was such that it appeared to be in excess of the threshold limit value (TLV). Threshold limit values refer to airborne concentrations of substances and in the case of quartz the TLV is a time-weighted average concentration for a normal 8-hour workday or 40-hour workweek and represents conditions under which it is believed that nearly all workers may be repeatedly exposed day after day without adverse effect. The policy of the Health and Safety Commission (HSC) is, that exposure should be kept as low as is reasonably practicable and in any case within published standards e.g. TLVs.

In view of this situation, London Transport decided to carry out further investigations at other places, including crew positions on trains on four tube lines, and to consult the Medical Research Council's experts in this field. Because London Transport railwaymen do not spend all their working lives, for eight hours a day, in tunnels it was necessary, in interpreting the results of these additional tests, to adjust them to the time-weighted average concentration of dust to which train crews were actually exposed.

Before being adjusted, the results of the tests on trains in the tunnels showed that the respirable quartz was up to three times the TLV on the Bakerloo Line, around the TLV level on the Northern Line, below it on the Central Line, and well below it on the Victoria Line. These results are in line with other evidence; the Bakerloo Line, for example, was already known to produce greater brake block wear than other tube lines while, as mentioned earlier, Victoria Line trains have rheostatic braking, which does not depend on friction (except at very low speeds).

After adjustment to allow for the average periods which train crews actually have to spend in the tunnels each day, the quartz exposure figures were much reduced; those for the Bakerloo Line still exceeded the TLV somewhat, but others were only a fraction of that threshold level.

The following diagram shows the average periods of time which train crews are booked to spend actually in tunnel in a shift, on each of the deep-level tube lines of the Underground system. These are the figures used in making the adjustments to the quartz exposure levels, as explained above.

These are averages. The maximum and minimum times were made available in the detailed supporting data. The Bakerloo figure of 4h 26 min is the time which applied when the original investigation was made. Since the opening of the Jubilee Line in May 1979 there have been integrated crewing arrangements between the two lines and the Bakerloo/Jubilee figure is 3 h 45 min.

The calculations described above suggest that, even on a strict interpretation of the standards, possible health hazards from quartz in Underground dust are small and likely to be limited to only one line.

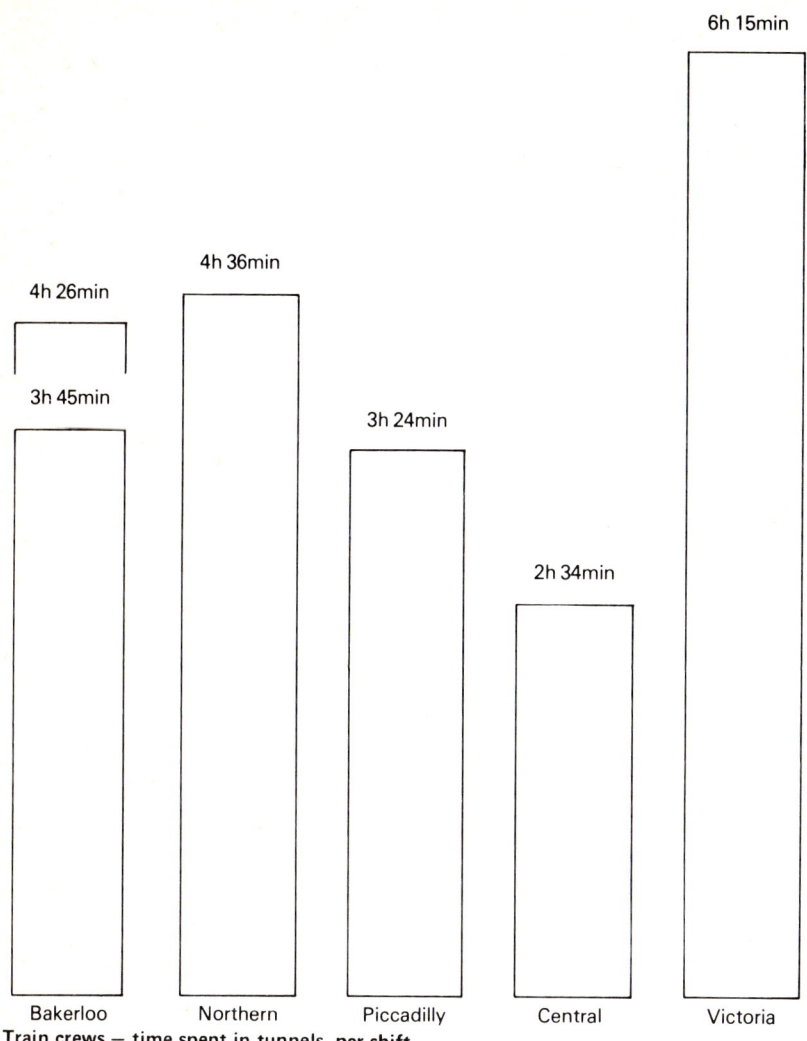

Train crews — time spent in tunnels, **per shift**
Average times, Monday—Friday
Fig 2

However, when London Transport sought the advice of the Medical Research Council's special unit for dust disease of the lungs (the Pneumoconiosis Unit) in 1976, the Unit's Director considered that because of the large amount of iron with which it was mixed the quartz in the Underground tunnel air might act differently from the quartz on which the the TLV standard was based.

The Medical Research Council was then asked to co-ordinate a three-pronged research project to measure the biological activity of the quartz in respirable dust. Firstly, an electrical test on the tunnel dust confirmed that the quartz had become, in effect, covered with the iron. Secondly, living cells were exposed to tunnel dust in the laboratory. Thirdly, respirable tunnel dust was instilled into the lungs of experimental rats. The results of these investigations indicated that tunnel dust produced only such changes as could be expected from 'nuisance' dusts which do not produce progressive fibrosis in animal inhalation studies and are not known to produce disabling pneumoconiosis in humans.

At the same time as these different research investigations were put in hand, London Transport carried out a special review of its medical records of train crews to see if there was any evidence of any harmful effects of tunnel dust on them. In the first place it was established that there was no record of any case of dust disease (pneumoconiosis) among train crews due to work in London Transport. Next the chest X-rays which form part of London Transport's many routine age examinations for Underground motormen were studied. The chest X-rays of the last 81 train crew staff examined who had worked for more than 15 years on tube lines, in most cases on the Bakerloo and Northern Lines, were taken out, and mixed at random with similar X-rays of 47 bus drivers which had been taken during routine medical examinations over the same period of time. All these X-rays, which were not labelled to show which were of bus drivers and which were of motormen, were then passed to the MRC Pneumoconiosis Unit for scrutiny by three separate observers. The results showed no pattern of lung trouble for either group, and no evidence was found of any lung disease which could be attributed to quartz, or dust generally, in the X-rays of the train crews, despite regular exposure to tunnel conditions in the course of their 15 years or more of tube service.

Attention was then turned to the records of sickness absences of four days or more caused by bronchitis, and again a comparison was made between train and bus staff. The result of this comparison over two periods, from 1968 to 1970 and from 1971 to 1976, showed that, on average, sickness absence due to bronchitis was actually lower among train crews than among male bus crews. The comparison for the second of the two periods is shown in the diagram on the following page, with a breakdown by age groups.

It is of interest that the medical records also show that whereas sickness absence due to bronchitis has increased among bus crews in recent years, it has remained fairly constant among train crews.

Two other results of the analysis of the sickness records should also be mentioned. The first is that, although, as shown earlier in this report, the Bakerloo Line appears to have the highest concentration of respirable quartz in the tunnel dust, the total rate of sickness absence for Bakerloo Line train crews corresponds closely to that for train crews generally throughout the Under-

Sickness absence due to bronchitis
Period 1971—1976
Fig 3

ground system. The second result, which is based on a Medical Research Council study of lung cancer and is to be published shortly in a preliminary report, shows that no excess of lung cancer has been seen among Underground motormen and guards, either in comparison with other London Transport male staff or in comparison with the general population of Greater London;

this conclusion is the outcome of analysing records over 20 years from 1950 onwards.

Altogether, from the variety of evidence quoted above from the tests of biological activity, staff chest X-rays and sickness records, there is no indication that the level of dust in London Transport tube tunnels and stations constitutes a hazard to staff or the travelling public.

Research methods

Before going on to consider what action has been and is being taken as a result of the research work described in this report, it is only right to show that the research methods used have been sound and that the findings are therefore not open to dispute.

Most of the research has been carried out by London Transport's own Research Laboratory, using the accepted standard equipment and techniques, but where more specialised facilities or advanced methods have been needed, London Transport has not hesitated to seek expert advice from the best qualified outside research organisations or ask them to carry out research studies themselves on London Transport's behalf.

For the measurement of total and respirable dust in the Underground the London Transport Research Laboratory employs standard (Rotheroe and Mitchell) portable samplers, as well as high-volume samplers where appropriate. At some individual sites where sampling is taking place cross-checks have been made at different positions to ensure that there is no 'freak' factor in the quoted results.

So far as asbestos is concerned, the earlier tests carried out in the 1950s by the London School of Hygiene used techniques which were crude by modern standards. Later, with the growth of knowledge on asbestos, the Asbestosis Research Council developed an improved detailed method for asbestos measurements, which has now been adopted to meet the legal requirements; this was the method used by the London Transport Research Laboratory in its more recent tests, using optical microscopes. To ensure reliable results, the laboratory staff were specially trained in asbestos counting by reputable outside bodies and it was also laid down that two people should independently count any sample which was analysed. Even with these methods, however, it was found that, because the quantities of asbestos in Underground dust were so very small, absolutely precise figures could not be obtained. So when it was learned that the Asbestosis Research Council had devised a very sophisticated method, using electron microscopes, to measure extremely small amounts of asbestos in the air, London Transport asked that Council (jointly with the TUC Centenary Institute of Occupational Health, attached to the London School of Hygiene) to measure the airborne asbestos concentration. As noted earlier in this report, the results gave further reassurance about the very low health hazard from airborne asbestos in the Underground.

As regards the quartz content of the respirable dust, this has been measured by a standard infra-red technique. For purposes of comparison, the London Transport Laboratory was supplied with standard samples of quartz-containing dust by the Health and Safety Executive (HSE). In the course of its work on quartz measurement, the Laboratory has checked its methods against those of other bodies engaged in the same work.

Further action

Apart from the biological experiments described earlier, a number of other lines of action are being followed which affect the tunnel dust position. These further developments are necessary, despite the very satisfactory outcome of all the research so far, because

(a)   it is still desirable to cut down the amount of dust in the Underground, if only from the standpoint of cleaner working and travelling conditions;

(b)   the staff and passengers will need continued assurances that any possible harmful substances in the Underground dust are being kept to a minimum, even though they are already within the present limits.

Reference has already been made earlier in this report to the combination of steps taken at the worst dust 'black spot' of all (Highgate), which reduced the amount of dust in the air there by nearly 85% in eight years — and to the signalling changes which reduced dust at Baker Street (Bakerloo Line) by 66% in fifteen months.

Since most of the dust problem stems from brake block wear, much of the effort has gone into the question of braking. Rheostatic braking is undoubtedly the most effective way of reducing the production of dust; as an illustration, analysis showed that the concentration of respirable quartz in the Victoria Line (which has rheostatic braking) was less than one-fifth of that in the Bakerloo Line (with conventional braking). It is now London Transport's policy that all new Underground rolling stock shall be designed with rheostatic brakes. In the particular case of the Bakerloo Line, there should be a marked improvement in a few years' time, when the existing old (1938) trains are withdrawn for scrapping and trains with rheostatic brakes are transferred to the line.

Action is also being taken on the composition of the brake blocks. In 1977 London Transport approached its two brake block manufacturers with a request that they should develop a new type of block containing much less quartz or no quartz. Financial help was offered to the manufacturers for this purpose, but the offer was declined for commercial reasons. One of the manufacturers produced a brake block containing only one-third as much quartz as the standard block. This block, which proved satisfactory in friction trials on a test locomotive and on trains, was introduced on the Bakerloo and Northern Lines. Longer term experience, however, showed that it was unsatisfactory in that it caused damage to the wheels, and eventually for that reason it had to be withdrawn. Preliminary test work is in hand with an alternative block subsequently put forward by that manufacturer. The second manufacturer developed a brake block which showed promise in preliminary trials, but there would have been technical difficulties in its general introduction. This manufacturer has now developed a block which is undergoing trial in service on the Bakerloo Line. Test programmes in conjunction with these and other manufacturers are being pressed ahead as quickly as possible.

London Transport has asked that the new experimental brake blocks should, if technically possible, also be free from asbestos. Indeed, despite the infinitesimal amount of asbestos revealed by the research programme in the Underground dust, it is now the policy of London Transport to avoid the use of asbestos wherever possible in its equipment, rolling stock and architectural works. As an example of this policy, the installation of asbestos-braided cables is being discontinued as soon as the new types of cable, which incorporate materials which reduce the fire hazard to such an extent that asbestos braiding is no longer necessary, become fully available.

Other lines of action include the purchase of new equipment, such as a scanning electron microscope, for the London Transport Research Laboratory, and the development of a new technique for quartz measurement, using the Laboratory's existing X-ray diffraction equipment. In addition the Laboratory has become a member of the Respirable Dust Analysis Group of the HSE, which exists to improve research methods in this field; and, as mentioned earlier, London Transport has obtained standard quartz dust samples from the HSE. These various steps will greatly assist the Laboratory in its continued checking on tunnel dust concentrations and contents.

Finally under this heading reference must be made to the Tunnel Cleaning Train which, as mentioned earlier, London Transport has been developing in recent years. It is important to understand the original purpose for which this train was designed, namely to remove accumulations of *settled* dust from the Underground tunnels. This in turn will reduce the unpleasantness which the settled dust creates for Underground maintenance staff, and will also cut down the risk of smoulderings in tube tunnels. To carry out its purpose the Tunnel Cleaning Train will have to pass large volumes of air through its filters, which must therefore be too coarse to retain all the finest particles; this means that the train will not be able to remove all the *airborne* dust in the respirable size range. In fact, in terms of quantities of airborne dust, these are likely to increase temporarily where the Tunnel Cleaning Train is used; but in any event, the train can deal with only a small length of tunnel at a time, and air moves freely throughout the system. In testing the Tunnel Cleaning Train, however, the opportunity is being taken to examine tunnel dust levels in a section of tunnel both before and immediately after cleaning.

Collaboration and communication

As already stated, London Transport has sought the most expert advice and help available whenever its own knowledge or facilities for investigating the dust problem have been insufficient. On other occasions, the official bodies responsible for applying the health and safety regulations have themselves carried out tests. The expert organisations which have been involved in one way or another are:

Asbestosis Research Council
Cardiff University
Health and Safety Executive, including
    HM Factory Inspectorate
    Industrial Hygiene Unit
    Railway Inspectorate
    Respirable Dust Analysis Group
London School of Hygiene and Tropical Medicine
McCrone Research Associates
Medical Research Council, including
    Pneumoconiosis Unit
TBA Industrial Products (Research & Engineering Division)
TUC Centenary Institute of Occupational Health
Winton Laboratories
Yarsley Technical Centre, including
    Yarsley Testing Laboratories

So far as contact and communication with London Transport's own Underground staff is concerned, there has of course been a continuing exchange between the management and the trade unions and staff representatives on the dust problem for many years. The staff have naturally been concerned about the effect of dust on their working environment and the possibility that some of the dust might present a health hazard. Staff representatives and the trade unions have therefore urged London Transport to press on with its researches and to keep them informed of the results. In consequence, the unions concerned are aware of virtually all the information contained in this report and of much of the more detailed supporting evidence. Also, on a few occasions, one of the unions has asked the Factory Inspectorate or Railway Inspectorate to carry out investigations independently, and in each case the results have been reassuring and in line with what London Transport would have expected.

There are various channels for regular discussion on this subject with the unions, including meetings at Sectional Council and 'Clause 6' (Union Executive and London Transport Chief Officer) levels; members of London Transport's Medical and Scientific Staff including on occasion the Chief Medical Officer and Scientific Adviser have attended these meetings and taken part in the discussions. On at least one occasion these officers have also attended a union branch meeting to discuss dust problems with the branch members. The Chief Medical Officer also gives a regular talk, dealing among other things with the dust question, to courses for Safety Representatives appointed by the unions under the Health and Safety at Work etc. Act 1974 (HSW Act). Also, information about particular developments is given out in the Traffic Circular, which is seen by all the operating staff; for example, notices were published about the removal of asbestos at the 'Bull and Bush' site on the Northern Line and the tests of airborne dust there after the work had been done. The trade unions are given the results of laboratory tests.

There are occasional complaints about dust in the Underground from local authorities, residents' associations and private individuals, sometimes in the form of letters to newspapers. Every effort is made to give full replies, based on the latest information available. Also, qualified London Transport Officers have taken part in radio and television programmes when the question of dust in the Underground has been discussed.

London Transport has naturally consulted other city transport undertakings on the subject of tunnel dust, but their conditions are different from those on the Underground, which is the only major 'Metro' system in the world to have small-diameter single-line tube tunnels with the minimum of clearance between the trains and the tunnel linings. Apart from a limited amount of work which the New York City Transit Authority has carried out on dust as part of a general survey of air conditions in the New York 'Subway', there appears to be no other special research going on. The International Union of Public Transport (UITP) has, however, circulated a report on asbestos on railways generally, which is in line with London Transport's own findings on the subject.

Summary

To sum up:

(a) At the dust 'black spots' on the Underground the total concentration of dust in the air has been dramatically reduced by enforcing speed restrictions, altering signalling layouts, and other measures. The position will be further improved by the introduction of more and more new rolling stock with 'rheostatic' (non-friction) brakes.

(b) Extensive research, supported by expert independent investigation and opinion, has shown that the concentrations of asbestos in the air in the Underground are far below the level at which there is any significant risk to health. The position will nevertheless continue to be monitored and London Transport will avoid using asbestos wherever possible in future.

(c) Extensive research has also shown that the possibility of health hazard from quartz in the air in the Underground is remote. The quartz levels are generally below the accepted limit. Moreover, the medical records of train crews covering many years show no signs of illness which could be attributed to exposure to quartz dust, or indeed any dust. Quartz dust levels will, however, continue to be monitored in the expectation that this will show that its production is being further reduced by the increased use of rheostatic (non-friction) brakes and, on existing trains, by the use of brake blocks with little or no quartz in them.

(d) No other constituent of Underground dust is suspected of causing any risk to health.

(e) London Transport's Tunnel Cleaning Train is designed to remove *settled* dust, thus relieving staff of dirty jobs and dirty conditions, and reducing the risk of smoulderings when it goes into service, but its effects on airborne tunnel dust will be subject to initial testing.

(f) The trade unions have been in discussion and consultation with the mangement of London Transport for many years on the question of dust in the Underground and are aware of all the research that has been carried out, the conclusions reached, and the action taken or being taken. In the proper interests of their members they have urged progress in dealing with this subject, and they will of course be advised of the results of the remaining tests and experiments as soon as these are available.

Whereas the main concern of this paper is with the possible effects of dust on train crews, station staff and the travelling public, parallel investigation of dust exposure of other staff working on the Underground system has taken place wherever it is thought necessary. In recent months, for example, there has been a joint HSE/London Transport investigation of the dust exposure of 'fluffers', who carry out duties of track bed sweeping. It has been necessary to monitor and control the dust encountered by staff and contractors carrying out the actual removal of sound-proofing insulation from tunnel walls, particularly in the Northern and Bakerloo line tunnels. In the latter cases the investigations have been carried out by London Transport's Research Laboratory and by an independent research laboratory. Railway Trade Unions and staff are kept fully informed of such work and copies of reports and results of any tests are communicated direct to them as well as being included in the agenda of any of the meetings which are held regularly or specifically for the exchange of information.

Supporting information

The detailed research results and other documents on which this report is based have been made available to the Railways Industry Advisory Committee and to the HSE.